Joécio Santos Sousa
Paulo Borges da Cunha

Urban environmental quality in the municipality of Teresina - PI

Joécio Santos Sousa
Paulo Borges da Cunha

Urban environmental quality in the municipality of Teresina - PI

Red and São Pedro neighbourhoods

SciencíaScripts

Imprint

Cover image: www.ingimage.com

This book is a translation from the original published under ISBN 978-613-9-69682-6.

Publisher:
Sciencia Scripts
is a trademark of
Dodo Books Indian Ocean Ltd. and OmniScriptum S.R.L publishing group

120 High Road, East Finchley, London, N2 9ED, United Kingdom
Str. Armeneasca 28/1, office 1, Chisinau MD-2012, Republic of Moldova, Europe
Printed at: see last page
ISBN: 978-620-8-13621-5

"Everyone has the right to an ecologically balanced environment, an asset of common use to the people and essential to a healthy quality of life... "

Art. 225, Brazilian Federal Constitution

SUMMARY

The city is an urban space in a constant process of morphological, structural, social, economic, cultural and environmental transformation, and it is these transformations that cause changes in the composition of the physical, political and organisational space of cities. Urban planning is a means of promoting a balance between the modifying components of the city and the community that inhabits it, seeking to ensure the quality of the environment. In urban landscapes, the human being is characterised as the main agent modifying the structures of the environment, seeking to adapt it to their basic needs such as: housing, schools, spaces for social and cultural practices, health, among others; to the extent that they are also the main element to be affected in the anthropic environment by these interferences. Environmental quality, for example, is directly linked to a population's quality of life, and it manifests itself through various elements that make up the socio-economic and environmental reality of urban space. In this sense, the environmental quality of cities changes as new orders of consumption for material goods and natural resources emerge. Therefore, the evolution of society and the development of new technologies intensify and reshape the socio-environmental panorama and consequently all aspects related to environmental quality. The city of Teresina, capital of the state of Piauí, has been undergoing a process of economic and property overvaluation on the national stage, and is therefore in the midst of a transformation of its space. For this reason, this study sought to measure the urban environmental quality of two neighbourhoods, Vermelha and São Pedro, both belonging to the central-southern region of the capital, which are also in the midst of real estate development. To do this, an experimental methodology was applied that relates the characteristics presented by the indicators listed to the population's procedures towards the environment, thus revealing neighbourhoods with a good urban environmental quality index.

Keywords: urban environmental quality; methodology; indicators; Teresina

SUMMARY

INTRODUCTION

In the current social context, a large number of people live in cities, basically more than half of the world's population. This reality reflects directly on the quality of the services that are provided for the well-being of society and consequently on the construction of its quality of life. However, what can be seen in urban agglomerations, whether small, medium or large, is the deterioration and deficiency of relations between human beings and between human society and the natural environment.

The challenge of building improvements to the urban ecosystem is intensified by **the "negative cyclical" process it** goes through, caused by the demand for and transformation of ecological factors (abiotic, biotic and anthropic environment) into goods that will not replace the energy and minerals taken from nature. Population demand in cities itself makes demands on natural resources and in response returns harmful impacts to the cycle.

Human-nature relations are manifested in the human instinct for survival and permanence of the species on planet Earth, even if in contemporary times this need is presented in a new guise. Profit, for example, is the dictator of **modern man's "economic"** survival**:** basically all natural and non-natural resources are commodities with added economic value and indispensable for maintaining the financial market. But it's worth emphasising that profit, from the exploitation of natural resources, is not in itself a villain, but rather its predatory and inconsequential action on the natural environment.

A clear reflection of the valorisation of economic development to the detriment of socio-environmental development is the planning of Brazilian cities, for example. In Brazil, the social-housing-environmental structures that accompany the construction of cities are concentrated, restricted and unequal, since on average the best living conditions are concentrated in the hands of small groups.

Social inequalities and rapid demographic growth, accompanied by inefficient public administration strategies, result in the fragmentation of urban space. This is reflected in housing occupation, as seen in Brazil's medium and large cities, favouring the demand for housing and their rapid growth, as well as **the process of "peripheralisation" of the urban fabric. However, the sprawl of cities** occurs in an irregular and disorganised manner, worsening the living conditions of the community that lives there. Basically, invasions occur in river valleys, environmental protection areas, hillsides and hilltops, among others, causing severe environmental degradation and quality of life problems.

Environmental quality is directly linked to a population's quality of life, and it manifests itself through various elements of the socio-economic and environmental reality of urban space. The environmental quality of cities changes as new demands for material goods and natural resources emerge. And it is precisely this demand that causes serious pressure on the environment, directly affecting society's well-being.

The interaction between the ecological factors that make up the urban ecosystem are used as a basis for transforming space, but the effects of human actions occur directly on them. One of the serious problems in cities is that they are planned without taking into account the fragile interrelationship between these factors.

When deficient, urban environmental quality not only interferes with the construction of society's quality of life, but also with the physical, structural and organisational environment of cities. The failure of public policies to improve people's lives is also hampered by the way in which the degree of complication of impacted environments is reflected, i.e. sometimes recovery actions fail due to the inefficiency of the methods used. However, there is a lack of technical staff and studies capable of devising strategies that can change this situation.

The challenge in measuring urban environmental quality comes up against the difficulty of constructing indicators that reflect reality without making the mistake of excluding indispensable elements for measurement. In agreement with Braga **et al** (2003, p.7), there are three main approaches to measuring urban environmental quality: firstly, biotic and abiotic, in the search for biological, physical-chemical or energetic indicators of the ecological balance of an ecosystem. Secondly, the economic aspect consists of monetary valuations of natural capital and the use of natural resources. And finally, the aspect that seeks to build indicators of sustainability and environmental quality that dialogue with aspects of the natural ecosystem and aspects of the economic system and the quality of human life, in other words, aspects related to the anthropic environment.

Observing the reality of environments can provide indicators capable of reflecting the three aspects mentioned by Braga et al (2003), as the city is a laboratory ready to be used. The study of everyday life in society serves as a basis for formulating precise indicators for measuring urban environmental quality. This is why an intimate relationship between the object of study and the researcher is essential.

In order to gain an understanding of Teresina's neighbourhoods and the infrastructural problems the municipality is experiencing, as well as the probable socio-environmental risks involving the population, this research sought to measure the environmental quality of the space, due to the relevant effect it has on the quality of life, and thus the well-being of all. To this end, the following specific objectives were established: to get to know the reality of the São Pedro and Vermelha neighbourhoods in the city of Teresina; to list the indicators that assess urban environmental quality; to develop an experimental methodology that measures urban environmental quality; to compare the levels of urban environmental quality in the neighbourhoods.

The following research aimed to use real indicators present in the reality of the neighbourhoods studied that together reflect urban environmental quality, the way in which relations between society and the natural environment take place, the challenges and what can be done to achieve a sustainable city. This initiative is an instrument for drawing up public policies for spatial planning, strategies for public administration and the inclusion of the population in city management, as well as a bibliographical reference to provide a basis and foundation for future studies on the urban environment and ecosystem and the relationship between them and human beings.

CHAPTER 1

THEORETICAL FRAMEWORK

The city is an urban space in a constant process of morphological, structural, social, economic, cultural and environmental transformation, and it is these transformations that cause changes in the composition of the physical, political and organisational space of city environments.

Planning is a means of promoting a balance between the city's modifying components and the community that inhabits it, seeking to ensure urban environmental quality and the population's quality of life.

Today, cities are home to more than half of the world's population, which means that most human actions are restricted to these spaces. Their social, economic, historical and cultural relations, among others, are present in the urban ecosystem. According to Carlos (1994) quoted by Cunha (2010, pg. 29), the city is an environment that involves various aspects of human daily life, such as [...] the way of living, thinking and feeling, **thus constituting a privileged "field" for class struggles and** social **movements of** all kinds, which question the standardisation of the city and urban life. As such, the city is the physical space in which human-human and human-nature relations are brought into conflict and harmony between them is sought through a long process of transformation and (self-)knowledge of oneself and the environment, as well as the individual needs of each person.

However, the transformations that cities undergo are a set of acts, plans and goals that seek to promote the well-being of the community, called urbanisation. **However, "the rapid urbanisation that Brazilian society underwent in the 20th century introduced** a new and dramatic meaning into the territory of cities: the injustices and inequalities of **society" (CUNHA, p.32). The terrible** reality in which Brazil finds itself is justified by poor urban planning or even the lack of it.

The environmental quality of cities is changing as new demands for material goods and natural resources emerge. And it is precisely this demand that causes serious pressure on the environment, directly affecting society's quality of life. According to LIMA (2007), environmental quality can be seen as a balance between the physical elements and the types of land use and occupation, reconciled with the various activities carried out in this space in order to promote the population's quality of life.

In urban landscapes, human beings are characterised as the main agents modifying the structures of the environment, seeking to adapt it to their basic needs, such as housing, schools, spaces for social and cultural practices, health, among others.

To the extent that it is also the main element to be affected in the anthropic environment by these interferences

Population swelling ends up generating a violent need for space and, with it, the disruption of the environment. The conflicts that arise will have a direct impact on urban structures, thus favouring various aspects that highlight the levels of socio-environmental conditions in the region.

Combined with these demands, poor or non-existent planning leads to the uneven distribution of urban facilities and adequate infrastructure to ensure environmental quality and quality of life, thus favouring the formation of fragmented social spaces.

1.1 Human-environment interaction

Since the dawn of time, the human race has been constantly influenced by variations in the external environment in all aspects of its history, be it cultural, social or physiological. As well as the reciprocal action given back to the environment.

According to Cavalheiro (1991) quoted by Moura (2010):

> activities resulting from human concentration cause a rupture in the functional structure of a natural environment. The result is anthropogenic drift - a term used by Monteiro (1978) to characterise the impact of human actions on natural systems. A new landscape thus emerges, derived from the natural landscape and altered by human activities.

The city, therefore, is the result of transformations of the natural environment according to the **cultural, social and organisational needs of human beings. "Who are the** main agents that make up the **urban** ecosystem" **(PAVEZZI** and SILVA, 2011, p. 2).

Also according to Pavezzi and Silva (2011), the energy needed to continue human activities is required too much from natural resources, mainly from non-renewable sources such as fossil fuels. In this way, the flows of energy and matter in urban ecosystems are substantially different from those in natural ecosystems.

Both the energy and matter that enter the urban ecosystem are transformed into consumer goods through the processing of raw materials, and when they are no longer needed or usable, they are disposed of in the form of solid waste, which is destined for areas located on the outer limits of urban spaces, such as landfills. And this transformation process, in turn, requires more demand for natural goods, causing a vicious cycle of exploitation and generating pressure on environmental resources.

Gómez (2004) also comments on the interrelationship between man and nature quoted by Berto (2008), ratifies that:

> The material exchange between man and nature is a process that takes place throughout human existence. As long as man exists, he will have to allocate part of society's time to appropriating the objects of nature and transforming them into objects for human use through labour. This is an insurmountable necessity of human reality. However, it is important to note that while it is true that man can never stop appropriating the objects of nature through labour, the way in which he does this is historically changing. Understanding the way in which men appropriate and transform nature is inextricably linked to the ways in which men relate to each other and the development of society's productive forces.

Man's appropriation of environmental resources will always be due to the eternal human need to overcome the limits of his abilities, since each generation creates new technologies that can make life easier in society and for the economic system.

However, these technologies began to cause an imbalance in nature's cyclical recovery capacity, so that what was extracted from the natural environment was greater than its capacity for restructuring. The main process resulting from the transformation of nature due to human cultural, political and economic needs is the formation of cities and the organisation of urban space.

1.1.1 Urban space and its characteristics

Correa (1995) defines urban space as the set of different uses given to land, where such uses define areas such as: the city centre, the location of areas destined for commerce, industry, housing, services and well-being and leisure, as well as those destined for the expansion of the urban network.

In agreement with Correa (1995), urban space is made up of social agents such as the owners of the means of production, landowners, property owners, the state and excluded social groups. Capitalist urban space is a social product, the result of actions that accumulate over time and add to a society's cultural, social and political concepts, making it an articulated fragmented space that directly reflects the social condition of the agents that make up the urban environment, their symbols and conflicts.

According to Santos (2008):

> It is in this fragmented and articulated space that the city emerges as a pole of attraction. Its benefits and attractions go beyond the simple desire to improve life and social wellbeing, but at the same time the city is repulsive and exclusionary, because urbanisation shows us the cruel side of space when we see the speculation that comes from the groups that own the land and capital.

Because of the scale of their activities, the owners of the means of production consume a large amount of space and, in parallel, their actions shape the city, directly interfering in the location of other land uses.

Landowners work to obtain the highest income from their properties and are interested in using them in the most remunerative way possible. The demand for land and housing depends on the emergence of new social classes with incomes capable of participating in the land and housing market.

Property owners are agents who carry out, partially or totally, the following operations: development; financing; technical study; construction or physical production of the property; and commercialisation or transformation of commodity capital into money capital, now plus profit.

The state acts in the spatial organisation of the city through the instruments it can use in relation to urban space, such as the right of expropriation, regulation of land use, public investment in the production of space, through drainage works, dismantling, landfills and the implementation of infrastructure, among others.

Socially excluded groups, in turn, are those who don't have the income to pay the rent for decent housing, let alone to buy a property. This is one of the factors that, alongside unemployment, illness and malnutrition, define the social situation of excluded groups. These people are left with the following places to live: tenements, improvised housing, housing estates provided by the state and slums.

The production of urban space is closely linked to the interplay of interests between its agents and participants, the result of the symbolic and contradictory relations of capitalism in its multiple facets (SANTOS, 2009).

As the aforementioned author emphasises, urban space is artificial, it is built in what was once a natural environment and then manipulated in a web of social actions, where the relationships between the actors involved will not always result in the applicability of solutions aimed at the wishes of the majority.

> On the other hand, it is worth emphasising that urban space is not tied to a mere social representation of space, as it goes beyond the limits of the city and/or the growing urban centres, as it is possible to find urban characteristics in rural areas and vice versa. These phenomena, in this overlapping of areas, generate attractions for the reproduction of capital and the conditions for financial development, and ultimately the urbanisation of these areas (SANTOS, 2009).

1.1.2 The formation of cities and urban planning

According to Cassilha and Cassilha (2009, p. 09), cities in their formation process present themselves as a major challenge. In the words of the authors, this situation is evident "as the urban network grows in a disorganised and excessive manner, the demand for infrastructure increases". The main causes of the rapid increase in space are demographic pressure and property speculation, which increasingly demand solutions from the environment to meet society's needs.

Still in agreement with the aforementioned authors:

> Cities are mutant, growing and changing as a result of people's daily activities, each with its own form of appropriation, intervening in space in a wide variety of ways. As a result of these interventions, without the necessary measures being taken, problems begin to emerge everywhere.

According to Sobral (2004) quoted by Macedo and Rocha (2010, p.3), cities are the biggest drivers of environmental impacts due to their rapid demographic and socio-spatial growth, since they are also subject to the physical processes that operate on nature, but with a different dynamic.

According to Mota (1999, p. 22), planning should be based on the concept of sustainable development, which is understood as meeting the needs of the present without compromising the ability of future generations to meet their own needs.

1.1.3 Brief consideration of the formation of the city of Teresina

Teresina was founded on 16 August 1852 and is the only capital in the northeast to be located inland. It was planned and built especially for the purpose of being the seat of state government, as it had better communication conditions with other centres and the coast than the old capital Oeiras (MELO and BRUNA, 2009).

It was built on the Corisco plateau at an altitude of 72 metres. As a planned city, it has a chequered layout with straight, wide streets and tree-lined squares. Based on this checkerboard plan, its streets were built no wider than 7 metres and its blocks no longer than 100 metres, with Praça da Constituição, now Deodoro da Fonseca, as its centre. Until the 1900s, the city expanded according to these guidelines, with Avenida Frei Serafim as its central axis, dividing it into north and south (VIANA, 2005).

The growth of Teresina took place in both a northerly and southerly direction, with the northerly direction being conditioned by natural obstacles, such as the meeting of the rivers. As a result, there was only space in the southern direction - the eastern direction was opposed by the Poti River (LOPES **et al**, 2009). According to Façanha (1998), in the following decades, urbanisation in Piauí gained new momentum and Teresina became the state's main city.

In the 1970s and 1980s, after the construction of the concrete bridge over the Poti River in the 1960s, which allowed the city to advance towards the east and southeast of the capital, associated with the installation of the Federal University of Piauí and the horse racing track beyond the river's edge. As the population left their homes in the central area to occupy the expanding peripheral neighbourhoods, the landscape of the centre changed and underwent significant transformations (OLIVEIRA **et al**, 2012).

1.2 Urban Environmental Quality

Urban environmental quality is the balance that exists between the three ecological factors that make up the environment (abiotic, biotic and anthropic). However, this balance requires that all forms of energy subtracted from and added to the environment are consumed and transformed without causing any damage or negative impact to the physical, biological and human environments.

The concept of urban environmental quality is intrinsically linked to quality of life, and vice versa. As stated by Luengo (1998) cited by Dias **et al** (2011), environmental quality is related to the ideal conditions of habitable space, in terms of comfort related to environmental, biological, economic, productive, socio-cultural, technological and aesthetic aspects in its spatial dimension.

It is essential that the process of measuring the urban environmental quality of a place can go through the most varied axes that juxtapose the environmental theme, in order to develop a holistic vision in relation to the reality of the facts.

However, there are authors and methodologies that prioritise a thematic axis to ascertain the quality of life and environmental conditions of an urban space. For example, Lombardo (1985) cited by Cunha (2010) emphasises issues related to natural aspects and does not stress socio-economic issues. However, these restrictions seek to investigate the problem through one of the biases of measurement methods, which does not imply that they are wrong.

According to Machado (1997) quoted by Gomes and Soares (2004), environmental quality standards vary between cities and the countryside, between cities in different countries or in the same country, as well as between areas of the same city. These variations in standards and characteristics are justified by the socio-spatial realities of each environment. Each city has its own specific infrastructural, technological and organisational characteristics.

1.2.3 The formation of indicators and the challenge of measuring them

Each environment has its own specific characteristics and the indicators chosen to quantify and measure its environmental quality must be appropriate to its reality. Dahl (1997) states that the biggest challenge for indicators is to provide a picture of the sustainability situation in a simple way, despite the uncertainty and complexity. The author also emphasises the differences between countries, the issue of cultural diversity and the different degrees of development as important factors in the construction of indicators.

Over the last few decades, the formulation of indicators has been consolidated as an important tool for planning and evaluating public policies, including urban environmental policy. The correct use and reading of indicators enables decisions to be strengthened, facilitating, among other dynamics, the participation of society.

Guimarães (2004), in his dissertation on the methodological contribution to the assessment of urban environmental quality from a cultural perspective, defines an indicator as: parameters, or a value calculated from parameters, providing indications about, or describing the state of, a phenomenon, the environment or a geographical area, of a greater amplitude than the information directly linked to the value of a parameter.

In this way, indicators seek to generate information about physical, political, biological, social and cultural characteristics, among others, in order to generate, through an evaluation process, the identity of a particular object of study, be it a neighbourhood, a city or a country.

As Cavalcanti (2000), quoted by Guimarães, ratifies:

> The word indicator comes from the Latin *Indicare,* which means 'to indicate', 'to give directions', 'to estimate'. Indicators simplify complex and abstract phenomena into a quantifiable form, because the abrasive use of complex information can make decision-making difficult. Indicators are therefore used to represent an overall situation on a more complex scale in a more simplified way. In other words, the main role of indicators is to show a trend in a simplified way (about a complex reality), serving the purpose of assisting in decision-making, whether it be in drawing up public policies or monitoring these policies. Indicators can also be used to monitor a desired improvement in indices identified as below the desired level, or to maintain the indices considered desirable.

1.2.4 Discussions on Urban Environmental Quality: study experiences

In short, the indicators seek to bring together all the phenomena to the point where the complex reality can be demystified and simplified to a situation where environmental management guidelines can be established in favour of a balance between human society and the natural environment.

CHAPTER 2

METHODOLOGY

2.1 Study area

The municipality of Teresina, capital of the state of Piauí, is located according to the **geographical coordinates 05° 05' 12" latitude South of the equator and 42° 48' 42" longitude** West of Greenwich, at the confluence of the Poti and Parnaíba rivers (MAP 01). It has a total area of 1,756 km^2 , of which 242 km^2 is urban (SEMDEC, 2010). It is geographically bordered by the cities of União and José de Freitas to the north; Palmeirais, Monsenhor Gil, Nazária, **Demerval Lobão and Curralinhos to the south; Altos, Lagoa do Piauí and Pau d'Arco do Piauí to the east and** the city of Timon (State of Maranhão) to the west. According to the 2010 demographic census, it had a population of 814,230 inhabitants.

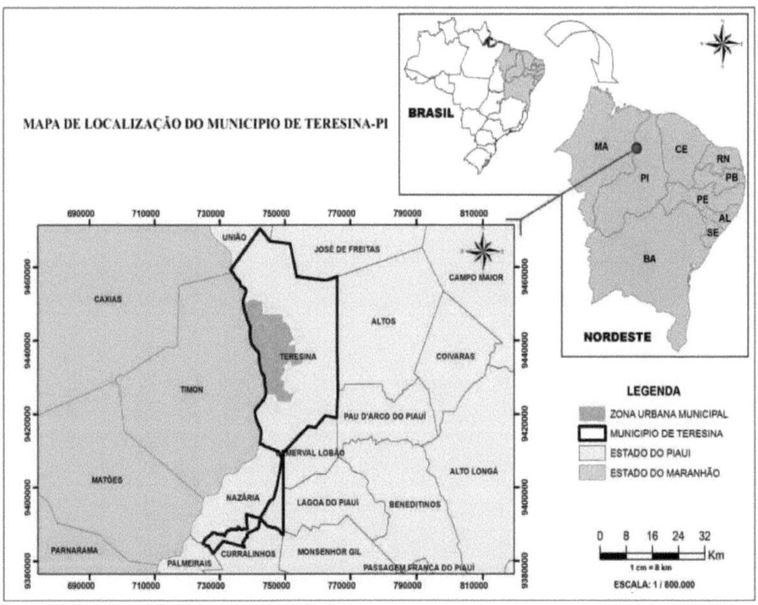

MAP 1- Location of the municipality of Teresina

Source: SOUSA, 2013

The town has one of the lowest altitudes in the state of Piauí. Located in a plateau area, it has flat terrain with gentle undulations (SEMDEC, 2010), and the average altitude of the municipality ranges from 100 to 150 metres and of its urban area is 72 metres.

Soil

The predominant soil types in the city are Yellow Lactosols and Red-Yellow Pedzolic soils, both of which have a medium texture. The former occurs more frequently in the flat portions of the municipality, notably in a strip parallel to the Parnaíba River, with a width of 10 kilometres, and in these areas they are deep and well-drained (CITYBRAZIL, 2013). Both types of soil have low fertility and marked acidity, resulting in soil with low nutrient content. In addition, reddish Brunizem soils, characterised by high clay content, low acidity and high nutrient content, are also present, albeit on a small scale (SEMDEC, 2010). Climate

The municipality of **Teresina is** popularly **known as the "hot city",** due to its location in a low latitude zone and on the edge of the semi-arid area of north-eastern Brazil. It has a tropical climate, one of the hottest in the country, and a sub-humid climate. Throughout the year the temperature varies between extremes of 22.0 °C and 40 °C. Due to its geographical position, the sun's rays hit Teresina with great intensity throughout the year, with the amount of radiation reaching the top of the atmosphere varying from 935 to 700 cal/cm2. From September to December, the highest temperatures occur, which is aggravated by various factors, such as deforestation and large-scale fires. The rainy season occurs in practically six months of the year, from December to May (SEMDEC, 2010).

Vegetation

According to the IBGE (2010), the city of Teresina has vegetation typical of the Caatinga and Cerrado biomes, the latter being predominant, represented by dense, medium-sized vegetation. In addition, according to SEMDEC (2010), the municipality has significant vegetation cover of forests and coconut groves (Palmeira do Babaçu), due to the influence of the neighbouring state (Maranhão).

The municipality is located in the great Parnaíba basin, which is permanently fed by groundwater from aquifers. It is also bathed by the Poti and Parnaíba rivers, which cover 53.75 km and 55.57 km of its surface respectively (SEMDEC, 2010). Considered the second most important river basin in northeast Brazil, the Parnaíba basin has an area of approximately 3,300,000 km^2 , covering 75% of the state of Piauí, 19% of the territory of Maranhão and 6% of the state of Ceará (LIMA cited by CUNHA, 2010).

2. 2Delimitation of the study area

The research took place in the São Pedro and Vermelha neighbourhoods, which make up the Central-South area of the city of Teresina. The neighbourhoods were chosen because of their representativeness in the group of neighbourhoods belonging to this region, because of the increasing value of real estate and because of the lack of studies on the subject in the area.

2.2. 1Red neighbourhood

Also part of the south-central neighbourhoods of Teresina, Vermelha (MAP 02) was named after the reddish colour of its soil. It was first called Quinta Vermelha (Red Farm), but over the course of its settlement it was recognised by the name it is known by today. It is bordered to the south by the

São Pedro and Pio XII neighbourhoods, separated by Rua Murilo Braga, to the north by Av. Joaquim Ribeiro (Centro), to the west by Av. Maranhão and to the east by the Macaúba neighbourhood, separated by Av. Miguel Rosa.

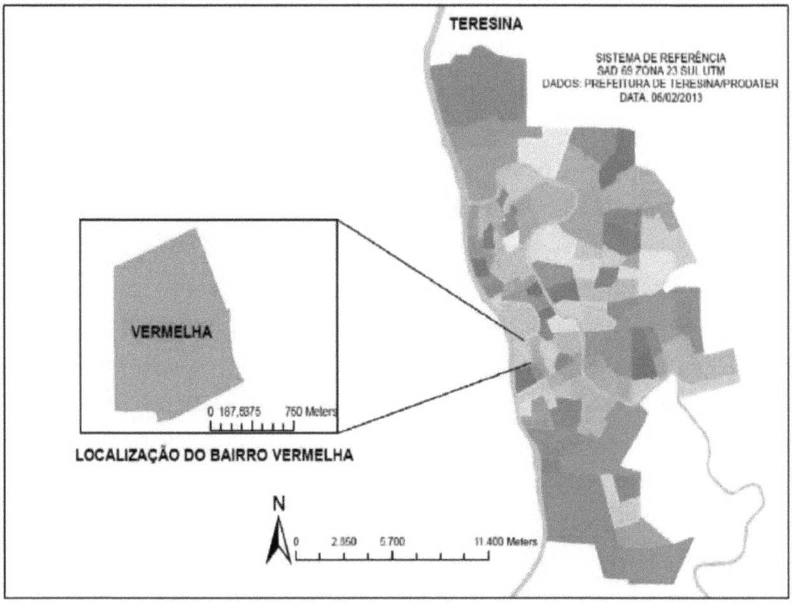

MAP 02 - Location of the Vermelha neighbourhood in the municipality of Teresina - PI

Source: Teresina City Hall/PRODATER -adapted by JOÉCIO SANTOS SOUSA, 2013.

2.2.1. 1Demographic aspects

a) Population

According to the 2010 demographic census, Vermelha had a total population of 5,659 inhabitants, of which 2,465 were men, corresponding to 43.55%, and 3,194 were women, corresponding to 56.45%. As can be seen from the population evolution of Vermelha, according to the results released by the 2010 census, it showed a drop in the number of its population, GRAPH 01.

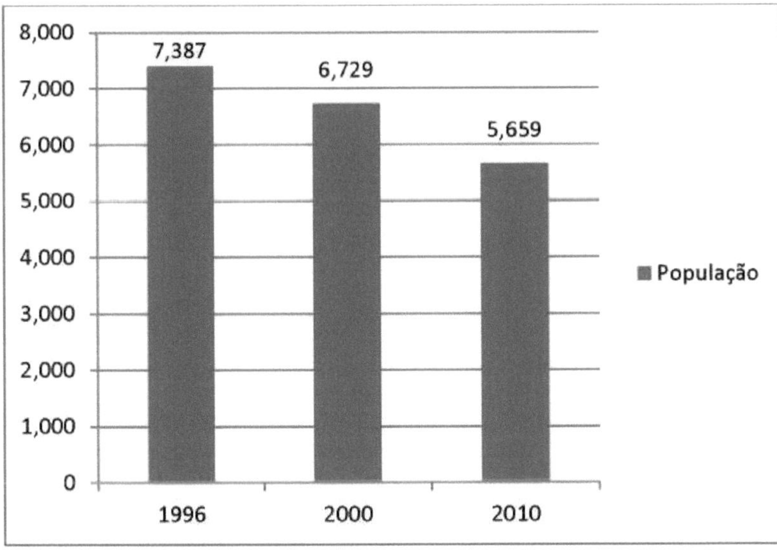

GRAPH 01: Population development in the Vermelha neighbourhood

Source: IBGE, 2010 - Adapted by JOÉCIO SANTOS SOUSA

2.2. 2São Pedro neighbourhood

The neighbourhood was first settled around the Catholic Church of São Pedro, hence the homage to the saint. Previously considered part of the Tabuleta and Vermelha neighbourhoods, São Pedro has gained an identity over the years. It is located in the Centre-South region of the municipality of Teresina (MAP 03). It is bordered to the north by the Vermelha neighbourhood, to the east by Pio XII, separated by Av. Barão de Gurguéia, to the south by the Tabuleta neighbourhood and extending as far as Av. Maranhão, which separates the neighbourhood from the left bank of the Parnaíba river to the west.

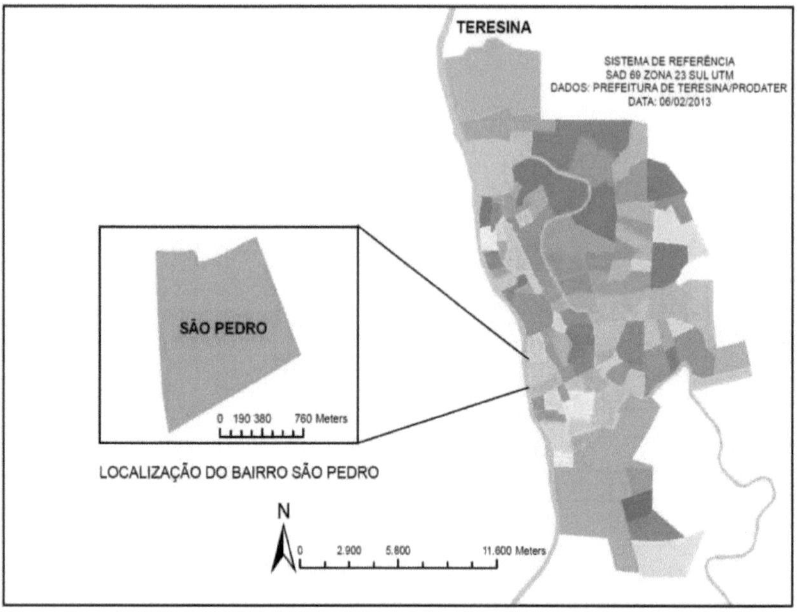

MAP 03- Location of the São Pedro neighbourhood in the municipality of Teresina-PI
Source: Teresina City Hall/PRODATER - adapted by JOÉCIO SANTOS SOUSA, 2013.

2.2.2. 1Demographic aspects

a) Population

According to the 2010 Census, the São Pedro neighbourhood has suffered a decrease in its total population in recent years (GRAPH 02). In 2010 it had a total of 8,780 inhabitants, of which 4,066 were men and 4,714 women.

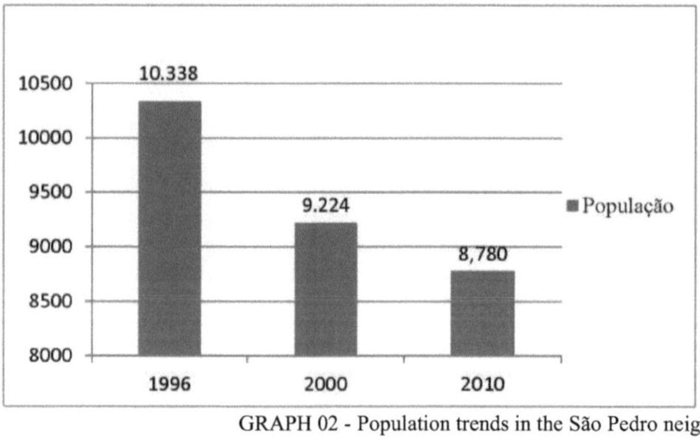

GRAPH 02 - Population trends in the São Pedro neighbourhood

Source: IBGE (2010) - adapted by JOÉCIO SANTOS SOUSA

2.3 Approaches , methods of data collection and analysis

The methodological procedures used in this work were applied in the municipality of

Teresina, capital of the state of Piauí, necessarily in the São Pedro and Vermelha neighbourhoods, which belong to the city's Centre-South region.

This is a quantitative-qualitative study, since it uses tools to quantify the elements that characterise the environments under study, both at the time of data collection and analysis, as well as social interaction to promote the collection of information, studying the particular phenomena of each environment in order to measure the environmental quality in which they find themselves.

As for the methods of approach, the research used the dialectical method, since it sought to study the phenomena that characterise the neighbourhoods based on the indicators listed, which influence the construction of urban environmental quality, as well as the reciprocal action between them and the dialectical changes they cause in nature and society. With regard to procedural methods, comparative and statistical methods were used, because in agreement with Marconi and Lakatos (2004), the comparative method seeks to ascertain the similarities and divergences of the sampling universe studied based on their comparison, and the statistical method **seeks to "reduce** sociological, political, economic etc. phenomena to quantitative terms and statistical manipulation, which allows the relationships **between** phenomena to be proven**, and generalisations to be made about their nature, occurrence and meaning".**

As a research technique, a structured form was used as a data collection mechanism, applied to the residents of both neighbourhoods, containing questions about environmental aspects; recurring environmental problems in relation to solid waste, sewage, green areas, noise pollution, air pollution, urban drainage, vector control; uses and origins with the quality of the environment and environmental perception.

2.3.1 Data collection

In order to obtain information about the urban environmental quality of the neighbourhoods studied, this research first had a theoretical basis in the literature and in discussions on the scientific scene in relation to the use of indicators that show the quality of urban spaces, seeking to adapt them to the reality of each location.

The theoretical basis for this work was built on research into scientific articles, doctoral theses and books on urban environmental quality, the construction of environmental indicators, the use of geoprocessing techniques to assess vegetation cover, among other relevant information. Basically, these references came from congress websites, scientific journals and academic institutions. The doctoral thesis was provided by Professor Dr Paulo Borges da Cunha (also the supervisor of this monograph), due to the closeness of the research problem and the interest in collaborating with studies on the subject in the city of Teresina. The books used were researched in the library of the Federal Institute of Piauí, Dr Francisco Montojos.

To characterise the neighbourhoods in terms of education, health, community associations, culture, sport and leisure, we used data published by the Municipal Planning Secretary - SEPLAM - on its official website. To characterise the neighbourhoods in terms of population, we visited the Brazilian Institute of Geography and Statistics (Instituto Brasileiro de Geografia e Estatística - Teresina unit) to obtain the data, which was updated according to the demographic survey carried out in 2010.

After gathering the relevant information for the theoretical basis, field visits were made to the neighbourhoods in order to get to know the reality of each location by observing the main problems and their origin in terms of solid waste, sewage disposal, green areas, leisure areas and sports.

2.3.1.1Construction of urban environmental quality indicators

In order to quantify the indicators that form part of this study, a photographic record was first made of the situations relating to the uses and origins of the urban space. Thus, each problem observed, such as inadequate solid waste disposal, open sewage, lack of public spaces for leisure and sport, was photographed for later consultation as to which indicators should be used to measure the level of urban environmental quality in the neighbourhoods, since it is extremely important to list indicators that are part of the reality and daily life of the environments.

> The indices that form part of this study seek to "construct indicators of sustainability and environmental quality that combine aspects of the natural ecosystem with economic aspects and the quality of human life, and in some cases also take into account aspects of the political, cultural and institutional systems" (BRAGA et al. pg 7).

These parameters tend to build a holistic view of urban problems, and their knowledge helps to promote management actions in favour of spatial planning.

In SCHEDULE 01 you can see the indicators of urban environmental quality according to their approaches, the physical environment: green areas, solid waste and urban infrastructure, and the anthropic environment: environmental perception.

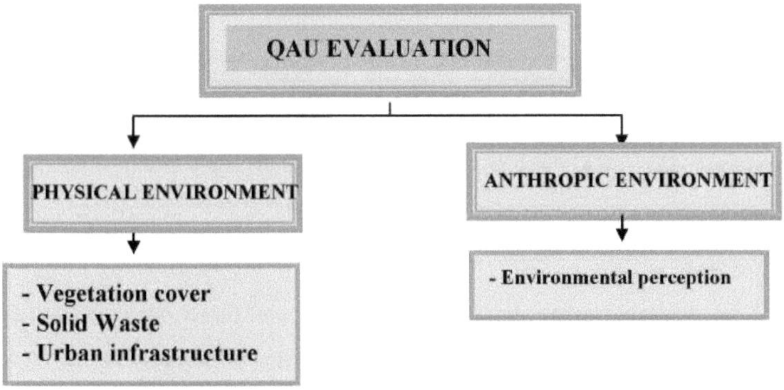

SCHEDULE 01 - Schematisation of approaches to measuring urban environmental quality

All the indicators mentioned here are part of the daily life of the neighbourhoods, in the construction and formation of their quality of life and environment. Since we have tried to combine all the possible indicators that make up the urban reality of these and other urbanised environments, determined on the basis of their approaches, as shown in the diagram above, they are directly linked to the community's actions on space, demonstrating how the quality of the relationship between man and the environment is expressed.

2.3.1.2 Application of forms with neighbourhood residents

The process of collecting data using a form, a model of which can be seen in APPENDIX I, was carried out in both neighbourhoods, but on different days. As already mentioned and also outlined in the previous topic, the form includes questions about environmental perception, environmental aspects and problems in the neighbourhood and at home, and the use and origin of the urban and home environment.

The first place to carry out this stage was the Vermelha neighbourhood, on 13 March 2013, from 3pm to 6pm. The procedure for choosing the streets was carried out by drawing lots, where the name of each street was written down, cut out and folded properly so that there was no privilege for one street over another when choosing. For the purposes of security, secrecy and ethics, this study does not reveal which streets were drawn, since it sought to measure the quality of the neighbourhood as a whole and not by fragments. Therefore, five streets were drawn.

To do this, a team of eleven members was set up, each of whom was tasked with applying one form per street, choosing alternate houses on opposite sides. In this first phase, the level of abstention was low and the population was receptive to answering the questions put to them. At the end of the day, a total of 55 forms had been applied in the Vermelha neighbourhood.

The second stage of data collection took place on 15 March 2013 in the São Pedro neighbourhood. The same methodology was used as in the first stage, but with a smaller number of participants. On this occasion, the similarities between the neighbourhoods could also be observed in terms of responses and willingness to collaborate with the research.

2.3. 2Data analysis procedures

The procedure for analysing urban environmental quality was based on observing the dynamics of the neighbourhoods in relation to the indicators presented in the subtopics below. To do this, an experimental methodology was devised in which negative and positive weights were assigned to each characteristic observed, plus its repetition along the neighbourhood boundaries. The higher the accumulation of positive scores, the higher the urban environmental quality; consequently, the higher the negative scores, the lower the neighbourhood's environmental quality (TABLE 01).

TABLE 01 - Urban Environmental Quality Index - *IQAU*

URBAN ENVIRONMENTAL QUALITY INDEX			
BAD	**REGULAR**	**GOOD**	**GREAT**
1 - 5	5 -10	10 - 15	15 - 20

The score for the indicator will be based on the following formula:

$$P_{ind} = p1 + p2 + p3 + ...pN$$

Where: P_{ind}: indicator score resulting from the sum of all weights

p: weight of the characteristic presented by the indicator

TABLE 02 shows the indicators of urban environmental quality, their characteristics and their weight.

TABLE 02 - Assignment of weights to urban environmental quality indicators

ITEM	INDICATORS	CHARACTERISTICS	WEIGHT
01	**Solid Waste**	proper storage and conditioning, regular collection	+
		improperly disposed of in spaces, abandoned on waste ground, outside the packaging container, burnt, abandoned in the public space	-
02	**Vegetation cover**	growth of vegetation cover	+
		decreasing vegetation cover	-
03	**Infrastructure and well-being**	existence of micro-drainage systems in a good state of repair, public spaces for sport, culture and leisure, pavements, health, education and social assistance facilities.	+
		lack of micro-drainage systems, health, education and social assistance facilities, lack of public spaces for sport, culture and leisure, pavements in a poor state of repair.	-
		greater perception of the dynamics of the neighbourhood	+

04	Environmental Perception	lower perception of neighbourhood dynamics	-

Once all the important data for the research had been collected, the process of analysing them began. For each indicator used here, a methodology was established to interpret and combine them in order to measure the environmental quality of each neighbourhood according to the weight established in the table above.

2.3.2. 1Solid waste

Solid waste is part of everyday life in any environment and when it is not managed properly, it causes serious environmental problems. In order to measure the environmental quality of the Vermelha and São Pedro neighbourhoods, solid waste was used as an indicator. This aspect was analysed according to its manifestation in the neighbourhood, its storage, packaging and disposal.

2.3.2.2 Vegetation cover

Using this indicator, this research sought to assess the evolution of changes in the green landscape of both neighbourhoods over two years using remote sensing technology, in which the technique used was the image classification process, the aim of which is to extract information in order to recognise patterns and homogeneous objects. The images were acquired on the website of the Institute for Space Research (INPE) from the LANDSAT 5 satellite for the years 2006 and 2010, both in June. In this case, the concept defined by Cavalheiro et al. (1999) cited by Dalbem and Nucci (2006) was used: vegetation cover is the projection of green on planimetric maps and can be identified using aerial photographs, without the aid of stereoscopy. The scale of the photo must accompany the vegetation cover indices; the location and configuration of the patches on maps must be taken into account.

In this case, the Vegetation Cover Index was obtained from the ratio between the area in hectares (ha) covered by vegetation and the total area of the neighbourhood, also in hectares, as shown in the formula below.

Formula:

$$ICV = Aveg / Atotal$$

Where:

ICV: Vegetation cover index

Aveg: Area covered by vegetation in a given neighbourhood in hectares

Atotal: Total area of the neighbourhood in hectares

For this purpose, the ideal situation was considered to be the parameter established by the United Nations (UN), which suggests a rate of 30% vegetation cover in urban areas, and if this figure is less than 5%, the region is characterised as an area similar to desert regions (Nucci 2008).

2.3.2.31 nfrastructure and well-being

In order to achieve a good quality of life or environmental quality, environments need to have adequate infrastructure so that residents can make use of their spaces, whether for walking, physical exercise, sports, travelling, living, etc. without damaging their physical and mental health.

Infrastructure and well-being were then used as indicators. This aspect looked at whether the micro-drainage system was in good or poor condition; the existence of sanitary sewerage coverage, paving, public water supply and lighting, health facilities, education, social assistance, spaces for physical exercise, leisure, culture and community socialising, such as squares.

2.3.2.32Environmental perception

In order to collect the environmental perception of the population of both neighbourhoods, a structured form was used with questions about the relationship between residents and the daily life of the neighbourhood and how they interact with the environmental changes resulting from urban dynamics, as well as their position and origin.

In this section, questions were asked about what they understood by the environment, environmental problems and how these affect them if they exist. These questions sought to extract information about what residents perceive around them, and how they influence the environmental quality of the neighbourhood represented by the sewage system, solid waste, tree planting and changes in the urban environment. Environmental perception encompasses the other themes discussed here and outlined in the following topics, so it provides a general overview of the situation in the neighbourhoods from the residents' point of view.

This indicator was assessed on the basis of the percentage of responses, so the higher the number of responses to the characteristics favourable to environmental perception, the more positive weight the environmental aspect will have, thus converging to a positive score at the end.

CHAPTER 3

RESULTS AND DISCUSSIONS

The sum of all the conflicts between anthropogenic actions and the natural environment is concentrated in the urban ecosystem. The rapid change that cities are undergoing is placing increasing demands on natural resources as a result of their formation and production process, since society is currently glimpsing the grandeur of its achievements without realising the real problems facing the environment.

Urban planning, for example, emerges in this theme as a mediator of the intrinsic relationships that exist between the three factors that make up the environment, namely the abiotic, biotic and anthropic environments. It seeks to shape space in order to reorganise it and adapt it to human needs without jeopardising the balance of other environmental factors. This results in favourable conditions for the community that lives there to enjoy a good quality of life.

However, as dynamic objects, cities need a periodic assessment of their structures, management models and, above all, the production of space, since human beings are constantly modifying their characteristics and causing impacted environments to emerge.

As a measure of the state of cities, indicators are used to help measure their level of quality, taking into account their structural elements.

3.1 Analysis of the Vermelha neighbourhood indicators

3.1.1 Solid Waste

When observing the neighbourhood in terms of where solid waste comes from, a large amount of improperly disposed of material was found in spaces such as roadbeds, near walls, on the public highway and outside the containers, as shown in PHOTO 01.

PHOTO 01 - Solid waste situation in the Vermelha neighbourhood
Source: SOUSA, 2013.

According to PHOTOGRAPH 01, we can see the residents' attitudes towards solid waste. Basically, it comes from household and economic activities. Photographs 1B, 1C and 1F stand out because of their peculiarities. The first shows the rubbish dump at Laurindo Veloso market (popularly known as Mercado da Vermelha), where you can see that the storage area is properly designed to contain the waste in a container and has an iron gate to keep the material safe. However, the dump was open, making it an attraction for vultures, who were scattering the solid waste on the street.

The second PHOTOGRAPH 1C shows one of the iron scrapyards in the neighbourhood, which particularly draws attention to the way the waste is stored and conditioned and its proximity to neighbouring homes. PHOTOGRAPH 02 below shows in detail the situation in which the waste was found.

PHOTO 02 - Scrap iron in the Vermelha neighbourhood
Source: JOÉCIO SANTOS SOUSA, 2013.

The solid waste is stored in an open area that is partly unprotected against violation by animals and people and/or the accumulation of water. Furthermore, it is located in a residential area, jeopardising the health of the area, as it is a breeding ground for insects and potential disease vectors (rats, cockroaches and dengue mosquitoes).

Finally, directly linked to the economic activities of automobile services and scrap metal, PHOTO 1F shows car carcasses disposed of on the roadbeds and pavements, aggravating the solid waste situation in the neighbourhood and potentially causing accidents for the community and passers-by. There is also evidence of the disposal of waste made up of potentially flammable substances (PHOTO 03), such as car lubricant oil, towelling and sponges contaminated by these components, as classified by ABNT NBR - 10004/2004.

PHOTO 03 - Solid waste from car services
Source: JOÉCIO SANTOS SOUSA, 2013.

Therefore, TABLE 03 shows the assessment of the indicator, solid waste, based on the characteristics they presented.

TABLE 03 - Assessment of the indicator - Solid Waste

ITEM	INDICATOR	FEATURES	WEIGHT
02	Solid Waste	Regular waste collection (three times a week)	+
		Red market waste exposed	
		Scrap iron with solid waste exposed	
		Car carcasses abandoned in the open	
		Disposal of potentially flammable material in the open	
TOTAL			- 4 points + 1 point

As a result of the assessment proposed in TABLE 03, the Vermelha neighbourhood, according to the Solid Waste indicator, scored a total of - 4 points, due to the sum of the negative weights assumed by its characteristics, and also scored + 1 point due to its positive characteristics. It is worth pointing out that this indicator alone cannot define urban environmental quality, since it is the result of the sum of all the indicators used in this research.

3.1.2 Vegetation cover

Remote sensing images from two different years (2006 and 2010) were used to assess vegetation cover, with a 4-year difference between the two. MAP 04 shows a negative change in the green patch in the Vermelha neighbourhood over the estimated period of time. TABLE 04 shows the percentage of vegetation cover and its loss in 2006 and 2010.

TABLE 04 - Percentage of vegetation cover in the Vermelha neighbourhood in 2006 and 2010

YEAR	PERCENTAGE	YEAR	PERCENTAGE	TOTAL LOSS
2006	25%	2010	17,50%	7,50%

As can be seen in TABLE 04, in 2006 Vermelha had 25% vegetation cover, which is considered by the Spring GIS to be any green spot. In MAP 04 for the same year, it can be seen that they are distributed throughout the neighbourhood, but mainly concentrated on the banks of the Parnaíba River.

In 2010, four years later, there was a drop in the rate of vegetation cover to 17.50%, totalling a loss of 7.50%, well below that established by the UN. Therefore, as shown in MAP 04, in 2010 the fragments of green patches diminished within the neighbourhood, mainly due to real estate speculation and the de-characterisation of vegetation.

As a result, we can see an increase in the residents' thermal sensation in relation to the neighbourhood, a situation that can be confirmed further on in TABLE 13 when, when it comes to the environmental aspect, tree cover, the residents point out that the little shade generated causes a more accentuated sensation of heat in relation to places with less vegetation cover.

With reference to the data presented above, TABLE 05 evaluates the indicator, vegetation cover, for the Vermelha neighbourhood.

TABLE 05 - Assessment of the indicator, vegetation cover - Vermelha neighbourhood

ITEM	INDICATOR	FEATURES	WEIGHTS
02	Vegetation cover	Diminishing vegetation cover in the neighbourhood	
		TOTAL	**- 1 point**

With regard to the result shown in TABLE 05, the vegetation cover indicator totalled - 1 negative point due to the decrease in green spaces in the neighbourhood over the estimated period of time between 2006 and 2010.

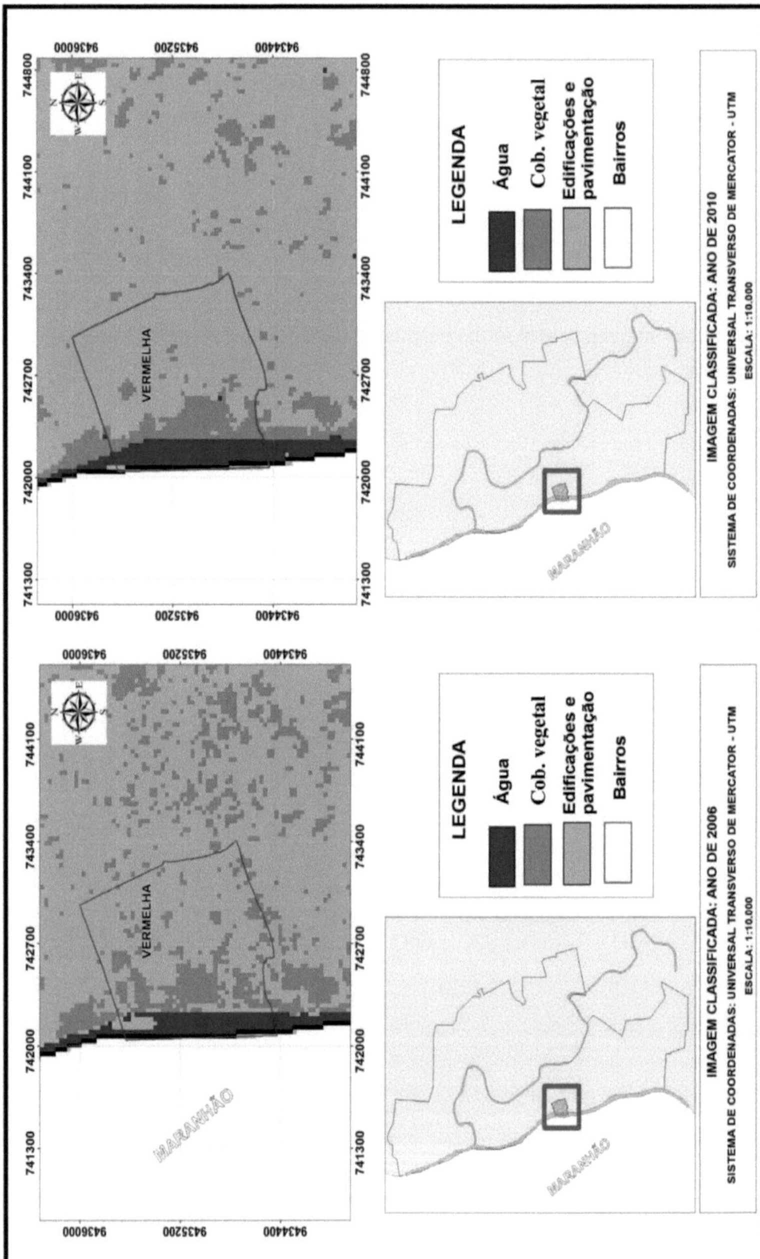

MAP 04 - Vegetation cover - Red neighbourhood
Source: SOUSA, 2013

3.1.4 Infrastructure and well-being

This indicator deals with urban infrastructures such as rainwater drainage, street structure, areas for sports, health, education, culture and leisure, which together make up the minimum quality of life needs of a population.

PHOTO 04 shows the structures of the urban microdrainage system found in the Vermelha neighbourhood

PHOTO 04 - Urban micro-drainage structures

Source: JOÉCIO SANTOS SOUSA, 2013

The basic objective of the urban microdrainage system is to guarantee the flow of rainwater, preventing the city from flooding. There are various functions of urban drainage systems, such as ensuring pedestrian and vehicle traffic, controlling erosion, protecting properties located in areas at risk of flooding and erosion, protecting public streets and roads, protecting and preserving the valley bottoms of watercourses and eliminating the proliferation of diseases in unhealthy areas.

Therefore, the urban microdrainage structures identified in the Vermelha neighbourhood, shown in PHOTOGRAPH 04, consist mainly of a culvert (4A), gutter (4B), and grilled manholes (4C and 4D). From the way they are presented, it can be seen that they are in a good state of repair and are not obstructed by solid waste.

The street structure is another factor used here to check the infrastructure and well-being indicator. In this case, attention was paid to the state of repair of the pavements. PHOTO 05 shows the street structure found in the neighbourhood and its deteriorated state of repair.

PHOTO 05 - Deteriorated street structure

Source: JOÉCIO SANTOS SOUSA, 2013.

As shown in PHOTOGRAPH 05, the street structure found in the Vermelha neighbourhood is in a poor state of repair, as it is bumpy and causes water to accumulate inside, making it difficult for passers-by and drivers to get around. Taking this fact as an example, during the application of the forms some of the interviewees commented on the structure of the neighbourhood's streets, claiming that this leads to falls, loss of footwear and even damage to car parts. However, it is important to note that the main avenues and streets in the neighbourhood have asphalt pavements, but most of them do not serve the areas where the houses are located.

Squares, gardens and parks are important mechanisms that make up the urban ecosystem for the purposes of contemplation due to their scenic beauty or for relaxation. They help to mitigate the local microclimate, making it more pleasant. In addition, they allow the population to use their spaces for group socialising as meeting points. As such, PHOTO 06 represents these spaces intended for outdoor socialising, such as the squares found in Vermelha.

PHOTO 06 - Squares - Red neighbourhood

Source: JOÉCIO SANTOS SOUSA, 2013

According to PHOTOGRAPH 06, you can see Praça da Saudade (6A) and Praça Nossa Senhora de Lourdes (6B).

The practice of sport and leisure are indispensable elements for building a good quality of life for a population, and for this to happen, the city needs to offer physical and structural conditions that encourage the practice of these activities. As such, PHOTOGRAPH 07 shows the leisure and sports facilities in the neighbourhood.

PHOTO 07 - Leisure and sports facilities in the Vermelha neighbourhood

SOURCE: JOÉCIO SANTOS SOUSA, 2013.

PHOTO 07 shows the football pitch and the senior citizens' gym.
In addition to these, Vermelha has a gymnasium integrated into the Gabriel Ferreira School Unit.

[3]In terms of culture, education, health and social assistance, the neighbourhood has 6 public schools (8A), the Church of Our Lady of Lourdes (8B), the Family Health Unit (8C) and the Guardianship Council (8D) (PHOTO 08), respectively.

PHOTO 08 - Infrastructure and well-being - Red neighbourhood

Source: JOÉCIO SANTOS SOUSA, 2013

So, after all the data on the infrastructure and well-being indicator has been analysed, TABLE 06 evaluates it based on the characteristics observed and expressed in the field.

TABLE 06 - Assessment of the indicator, Infrastructure and well-being

ITEM	INDICATOR	FEATURES		WEIGHT
04	**Infrastructure and well-being**	Urban drainage		+
		Street structure - paving		-
		Outdoor socialising - squares		+
		Sport and leisure	Football pitch	+
			Gymnasium	+
			Academy for the elderly	+
		Schools		+
		Family Health		+
		Social assistance - Guardianship Council		+
		Culture - Our Lady of Lourdes Church [3]		+
TOTAL				**+ 9 points** **- 1 point**

According to the assessment in TABLE 06, the infrastructure and well-being indicator totalled +9 points as a result of the sum of its positive characteristics, arising from the infrastructure and well-being equipment and instruments. However, because the pavement is in a state of deterioration, this indicator received - 1 negative point.

3.1.6 Environmental perception

Environmental perception is the way in which human beings perceive the events that surround them and are part of their daily lives. Basically, it is linked to the individual's level of awareness, to a greater or lesser extent, of environmental issues, the dynamism and transformation of space and the way in which it is managed, as well as changes and trends in behaviour.

The social profile of the population studied in the Vermelha neighbourhood was as follows: 54.5% women and 45.5% men. In terms of level of education, 3.45% were illiterate, 9.09% had completed primary school and 14.54% had incomplete primary school, 29.09% had completed secondary school and 9.09% had incomplete secondary school, 12.72% were in the process of completing higher education and 20% had completed higher education.

In order to extract the residents' environmental perception, they were asked what they understood by environmental problems, what type they observed in the neighbourhood and how they affected them, as well as other issues related to the aspects and dynamics of urban space. As a result, 80% of those interviewed said they observed some kind of environmental problem, which can be seen in TABLE 07.

TABLE 07 - Environmental problems and their consequences according to residents

ENVIRONMENTAL PROBLEM	OBSERVED CONSEQUENCE
Rubbish	Diseases; bad odour; clogs drains; proliferation of rodents and mosquitoes
Noise pollution	Disturbs concentration; disrupts sleep
Lack of Basic Sanitation	Bad odour; diseases
Lack of pavements	Pothole; causes fall
Public insecurity	Drug use; robberies
Insufficient afforestation	Increased heat
Air pollution	Respiratory problems; dirt

According to TABLE 07, various problems and their consequences could be listed according to the perception of the residents of Vermelha, but when discussing the environmental aspects that make up the neighbourhood and the relationship between these and their consequences, there were disagreements with the answers given, a situation that can be seen in TABLES 08, 09, 10, 11, 12 and 13.

TABLE 08 - Problems related to the environmental aspect - sewage system - according to residents of the Vermelha neighbourhood

ENVIRONMENTAL ASPECT OF THE NEIGHBOURHOOD	RELATED PROBLEMS	PERCENTAGE THE ANSWERS
Sewerage system	Exposed to the open air	32%
	It gets in the way	4%
	Accumulation of dirt	26%
	No problem	30%
	Others	8%

In line with the results above, 32% said they saw open sewage (PHOTO 09), due to the lack of a sewage system, which flows through the micro-drainage system. Another situation that also bothers residents is the accumulation of dirt as a result of the situation mentioned above, which appears at 26%. But 30 per cent said they didn't see any problems related to the sewage system, even if it was being dumped into gutters and storm drains.

PHOTOGRAPH 09 - Sewage exposed to the open air

Source: JOÉCIO SANTOS SOUSA, 2013

TABLE 09 - Consequence observed as a result of the environmental aspect - sewage system - according to residents of the Vermelha neighbourhood

ENVIRONMENTAL ASPECT OF THE NEIGHBOURHOOD	OBSERVED CONSEQUENCE	PERCENTAGE OF RESPONSES
Sewerage system	Insect and rodent proliferation	31,42%
	Exudes a bad odour	48,57%
	Diseases	31,42%

TABLE 09 shows that bad odour appeared as the main harm, with 48.57% of the responses. Diseases and the proliferation of insects and rodents came in at 31.42% each, which are the result of open sewage and the accumulation of dirt.

Another factor associated with the healthiness of the environment is solid waste, which has a direct impact on the health of the neighbourhood. Residents were therefore asked what kind of problems they observed as a result of solid waste and its consequences for their quality of life. TABLES 10 and 11 show the answers to the above problems, respectively.

TABLE 10 - Problems related to the environmental aspect - Solid waste - according to residents of the Vermelha neighbourhood

ENVIRONMENTAL ASPECT OF THE NEIGHBOURHOOD	RELATED PROBLEM	PERCENTAGE OF RESPONSES
Solid Waste	Insufficient collection	16%
	Proliferation of disease vectors	2%
	Stored improperly	28%
	No problem at all	58%
	Collected incorrectly	2%
	Burned	2%
	Abandoned in a vacant lot	2%

TABLE 10 shows that more than half of those interviewed said they didn't see any problems with solid waste, or any aspect related to it. On the other hand, 28% claimed that inadequate storage on pavements is the main difficulty. And 16% believe that the rubbish collection system should be more frequent, because the way it is (three times a week) is not enough to meet demand. The remainder totalled 2% each.

TABLE 11 - Observed consequences of the environmental aspect - Solid waste - according to residents of the Vermelha neighbourhood

ENVIRONMENTAL ASPECT OF THE NEIGHBOURHOOD	OBSERVED CONSEQUENCE	PERCENTAGE THE ANSWERS
Solid Waste	Diseases	23,68%
	Attracts animals	31,57%
	Bad odour	21,05%
	Impedes circulation	2,63%
	Visual pollution	21,05%

The ability to attract animals (PHOTO 10) is due to its storage and disposal conditions and this gave the environmental aspect, solid waste, the main consequence observed by residents, with 31.57 per cent of responses. This was followed by the alternative, diseases, which are the result of poor or non-existent management, with 23.68 per cent. The option bad odour, caused by the decomposition of waste, came in at 21.05%. Visual pollution and obstruction of traffic, with 21.05% and 2.63% respectively.

PHOTO 10 - Vultures attracted by solid waste exposed to the open air

Source: JOÉCIO SANTOS SOUSA, 2013

With regard to the characteristics related to the aspect of afforestation, residents were asked what problems they saw and their consequences, so TABLES 12 and 13 show the information obtained from the proposed situations.

TABLE 12 - Problems related to the environmental aspect - afforestation - according to residents of the Vermelha neighbourhood

ENVIRONMENTAL ASPECT OF THE NEIGHBOURHOOD	RELATED PROBLEM	PERCENTAGE OF RESPONSES
Afforestation	Insufficient	36%
	Poorly cared for	10%
	Planted in inappropriate places	4%
	No problem at all	56%

In accordance with TABLE 12, 56% of the answers given by residents of the Vermelha neighbourhood indicate that there are no problems with the trees in the area, and that they are satisfactory as they stand. However, 36% said that they were insufficient and did not meet the needs of the neighbourhood. The other alternatives, such as poorly cared for and planted in inappropriate places, came in at 10% and 4% respectively.

TABLE 13 - Observed consequences of the environmental aspect - afforestation - according to residents of the Vermelha neighbourhood

ENVIRONMENTAL ASPECT OF THE NEIGHBOURHOOD	OBSERVED CONSEQUENCE	PERCENTAGE OF RESPONSES
Afforestation	Little shade	81,81%
	Conflicts with infrastructure	18,18%

As a result of the dissatisfaction with the number of trees that appeared in the answers in TABLE 12, residents said that the main consequence of this is the little shade generated, thus increasing the thermal sensation of the environment. The conflict generated by the property or any urban infrastructure equipment was the second most mentioned consequence, with 18.18%.

Therefore, in order to bring together all the information provided above, TABLE 14 evaluates the percentage of responses in relation to the three environmental aspects listed on the residents' environmental perception of the dynamics of the neighbourhood.

TABLE 14 Assessment of the Environmental Perception indicator

ITEM	INDICATOR	FEATURES	WEIGHT
05	**Environmental Perception**	Perception of the environmental aspect - sewage system	+
		Perception of the environmental aspect - solid waste	-
		Perception of the environmental aspect - afforestation	-
TOTAL			**- 2 points**
			+1 point

Thus, TABLE 14 shows that of the three environmental aspects used to assess the environmental perception of the residents of the Vermelha neighbourhood, two obtained negative weights in relation to solid waste and urban afforestation, resulting in a final score of - 2 points. In addition to this, the indicator also scored +1 due to the population's perception of the environmental aspect of the sewage system.

3.2 Analysis of indicators in the São Pedro neighbourhood

3.2.1 Solid Waste

In the São Pedro neighbourhood, we can identify the different ways in which residents deal with solid waste, from its storage, disposal and destination. PHOTOGRAPHS 11 and 12 represent the situation and characteristics in which solid waste was found, but in the positive and negative orders, respectively.

PHOTO 11 - Positive situations and characteristics of solid waste
Source: JOÉCIO SANTOS SOUSA, 2013.

PHOTOGRAPH 11 shows the positive situations and characteristics of solid waste in the São Pedro neighbourhood. It shows waste being stored in closed bags (11A and 11C) and properly disposed of in containers (11B), organised without compromising the movement of passers-by. It's worth noting that during the visit to the neighbourhood, this behaviour was repeated several times.

However, it was also possible to notice the repetition of situations that added negative characteristics to the residents' approach to solid waste, as shown in PHOTO 12.

PHOTO 12 - Negative situations and characteristics of solid waste

Source: JOÉCIO SANTOS SOUSA, 2013.

As shown in PHOTO 12, there is inadequate disposal of solid waste on the roadbed and near walls, as in PHOTOGRAPHS 12A, 12B, 12E and 12F. In addition, PHOTO 12D shows the burning of waste near a vacant lot, which can cause the flames to spread to neighbouring homes. Also visible in PHOTO 12C is the accumulation of scrap iron on the wasteland.

As a result of the above situations, TABLE 15 assesses the score for the solid waste indicator, based on the weights assumed by the characteristics presented in the field.

TABLE 15 - Evaluation of the Solid Waste indicator

ITEM	INDICATOR	FEATURES	WEIGHTS
01	**Solid Waste**	Regular waste collection (three times a week)	+
		Proper storage	+
		Proper waste disposal in containers	+
		Solid waste abandoned on wasteland	-
		Burning solid waste	-
		Inadequate waste disposal on the roadbed	-
		Inadequate waste disposal near walls	-
TOTAL			**+ 3 points**
			- 4 points

In this way, the solid waste indicator, according to the assessment in TABLE 15, scored +3 points as a result of the sum of the positive weights assumed by the characteristics presented. However, the neighbourhood also had negative characteristics in relation to certain aspects of solid waste management, such as the inadequate disposal of waste on vacant lots, near walls, on the roadbed and in some cases associated with its combustion, resulting in - 4 points.

3.2.2 Vegetation cover

According to MAP 05, there has also been a decrease in green spaces in the São Pedro neighbourhood over the estimated period of time between 2006 and 2010. In recent years, the neighbourhood has been the target of real estate speculation due to the advent of new health and higher education facilities, which has led to an increase in demand for space to build housing and developments.

TABLE 16 shows the percentage of vegetation cover in the São Pedro neighbourhood in the estimated years and its loss over this time interval.

TABLE 16 - Percentage of vegetation cover in the São Pedro neighbourhood in 2006 and 2010

YEAR	PERCENTAGE	YEAR	PERCENTAGE	TOTAL LOSS
2006	50%	2010	27,50%	22,50%

According to TABLE 16, the São Pedro neighbourhood had 50% vegetation cover in 2006. It can also be seen in MAP 05 that, in this case, there is a greater concentration of green patches in the areas bordering the Parnaíba river. After four years, in 2010, the neighbourhood had an ICV of 27.50%, totalling a loss of 22.50% over this estimated period of time.

As can be seen from the data presented and MAP 05, this decrease is due to the accelerated growth of the other areas of the São Pedro neighbourhood and confirms the demand for space in the areas close to the new developments, such as the São Pedro hospital and the Santo Agostinho college, because MAP 05, 2010, shows a significant decrease in green spaces in the easternmost areas, close to Av. Barão de Gurguéia.

TABLE 17 evaluates the vegetation cover indicator for the São Pedro neighbourhood based on the characteristics presented in the discussion above and the data presented.

TABLE 17 - Assessment of the indicator, vegetation cover, São Pedro neighbourhood

ITEM	INDICATOR	FEATURES	WEIGHTS
02	Vegetation cover	Diminishing vegetation cover in the neighbourhood	
TOTAL			- 1 point

With regard to the result obtained in TABLE 17, the assessment of the indicator, vegetation cover, guaranteed the São Pedro neighbourhood - 1 negative point as a result of the decrease in vegetation cover in the neighbourhood during the estimated period of time between 2006 and 2010.

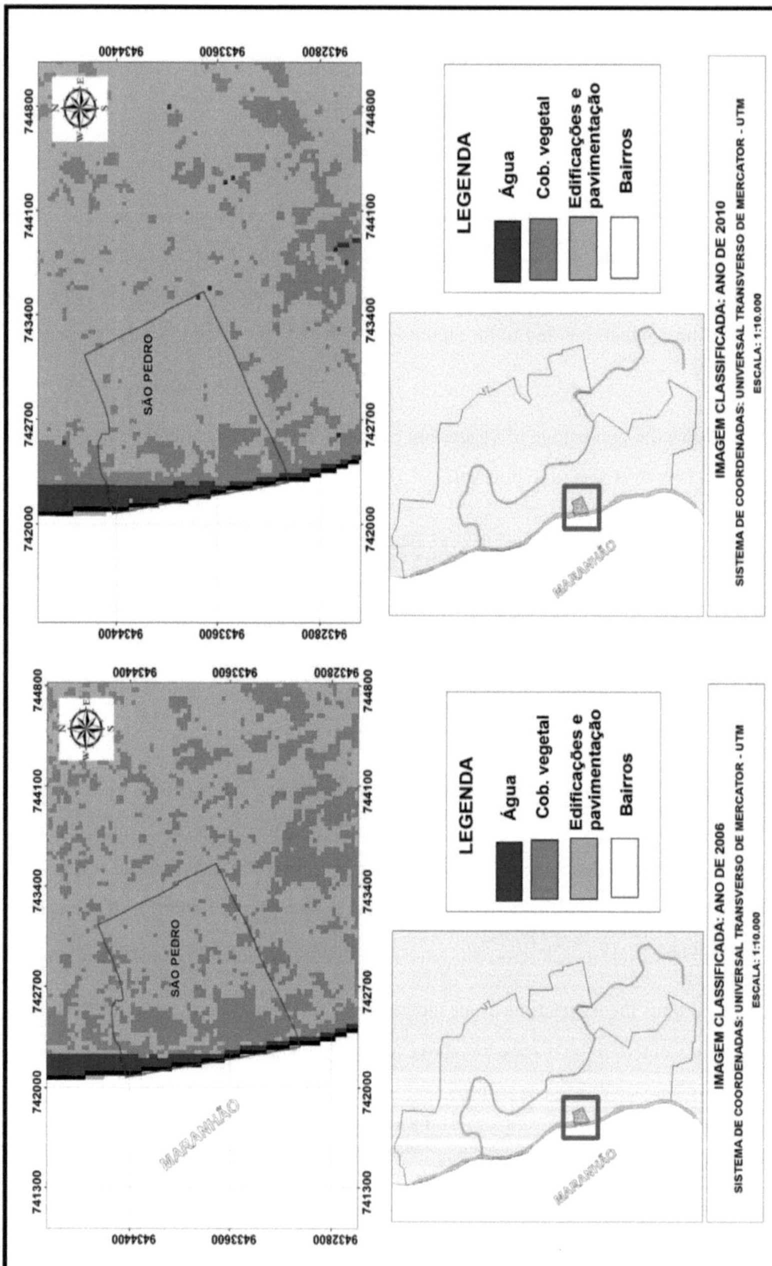

MAP 05 - Vegetation cover - São Pedro neighbourhood

Source: SOUSA, 2013

3.2.3Infrastructure and well-being

When evaluating the infrastructure and well-being indicator for the São Pedro neighbourhood, we looked at the condition of the urban rainwater microdrainage system, the state of conservation of the road pavement, areas for sports, culture, leisure, socialising, health and education.PHOTO 13 below shows the rainwater microdrainage system identified in the São Pedro neighbourhood.

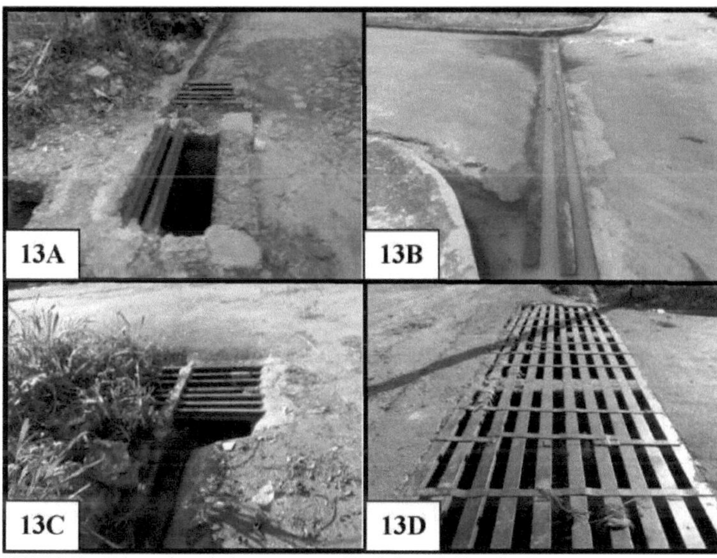

PHOTO 13 - Urban micro-drainage system - São Pedro neighbourhood
Source: JOÉCIO SANTOS SOUSA, 2013

According to what can be seen in PHOTOGRAPH 13, the rainwater drainage mechanisms found in the São Pedro neighbourhood are basically made up of gutters, culverts (13B), grilled manholes (13A and 13 C) and galleries (13D). In terms of what was identified, it can be seen that the mechanisms were not defective and were not even obstructed by solid waste or other material that would make it impossible for the water to drain freely.

As for the street structures, the situation was similar to the Vermelha neighbourhood, since only the main access streets and avenues have asphalt paving and the other roads were in a state of deterioration due to the potholes that had formed along the streets, as can be seen in PHOTO 14.

PHOTO 14 - Street structure - pavement - São Pedro neighbourhood

Source: JOÉCIO SANTOS SOUSA, 2013

As far as areas for culture, education, health and social assistance are concerned, there are several structures in the São Pedro neighbourhood, such as the São Pedro Church (15ª), the Digital Station (15B), 6 (six) school units, the São Pedro Health Centre, the Dr. Raul Bacellar Diagnostic Test Centre (15D), and the Piauí Association of the Blind - ACEP (15C).

PHOTO 15 - Infrastructure and well-being facilities - São Pedro neighbourhood

SourceJOÉCIO SANTOS SOUSA, 2013

As for squares, the São Pedro neighbourhood has Pracinha de São Pedro I and Pracinha de São Pedro II. As for spaces for practising sport, there is a sports court at the Association of the Blind.

In this way, TABLE 18 assesses the indicator, infrastructure and well-being, in relation to the characteristics it presents.

TABLE 18 - Assessment of the indicator, infrastructure and well-being, São Pedro neighbourhood

ITEM	INDICATOR	FEATURES	WEIGHT
		Urban drainage	+
		Street structure - paving	-

		Outdoor socialising - squares	+
		Sports court	+
		Schools	+
		Digital Station	+
		Healthcare establishments	+
04	**Infrastructure and well-being**	Social assistance - ACEP	+
		Culture - St Peter's Church	+
TOTAL			**+ 8 points**
			- 1 point

With regard to the results shown in TABLE 18, the São Pedro neighbourhood scored +8 points in relation to the sum of the positive weights presented by the indicator's characteristics, and also scored -1 point.

3.2.4 Environmental Perception

The sample population used to answer the forms in the São Pedro neighbourhood had the following social and educational profile: 50% women and 50% men. In terms of schooling: 4.54% have completed primary school and incomplete secondary school, 18.18% have incomplete primary school, 36.36% have completed secondary school, 22.72% have completed higher education and 13.63% are in the process of completing higher education.

On this occasion, questions were also asked to extract the environmental perception of the residents of the São Pedro neighbourhood, based on questions that dealt with aspects of daily life in the neighbourhood, the perception of the existence of environmental problems and how these influenced their lives. As a result, 81.81% said they observed some environmental problem in the neighbourhood and its likely consequences, a situation described in TABLE 19.

TABLE 19 - Environmental problems and their consequences according to residents of the São Pedro neighbourhood

ENVIRONMENTAL PROBLEM	OBSERVED CONSEQUENCE
Rubbish	Diseases, visual pollution, attracts animals, nuisance
Lack of basic sanitation	Bad odour, dirty street
Lack of tarmac	Causes accidents, breaks the car
Few trees	Heat

According to the data shown in TABLE 19, the main environmental problems pointed out by residents when the forms were applied in the field can be seen. These include rubbish, lack of basic sanitation, tarmac and insufficient tree planting. They also listed the consequences of these problems, such as visual pollution, bad odours, accidents and heat.

Other questions pertinent to the dynamics and transformation of the neighbourhood were raised during the interview, which questioned the environmental aspects that make up the urban ecosystem and which are also part of everyday life in the São Pedro neighbourhood, information about which is shown in TABLES 20, 21, 22, 23, 24 and 25.

TABLE 20 deals with the population's perception of the environmental aspect of the sewage system, and the probable problems that may arise in this area.

TABLE 20 - Problems related to the environmental aspect - sewage system - according to residents of the São Pedro neighbourhood

ENVIRONMENTAL ASPECT OF THE NEIGHBOURHOOD	RELATED PROBLEM	PERCENTAGE THE ANSWERS
Sewerage system	Exposed to the open air	40,70%
	It gets in the way	7,40%
	Accumulation of dirt	25,95%
	No problem at all	18,57%
	No sewage system	7,40%

According to TABLE 20, the main problem related to the environmental aspect of the sewage system is its exposure to the open air (PHOTO 16), which appeared in 40.70% of the responses. The accumulation of dirt was the second most frequently mentioned problem, with 25.95 per cent. On the other hand, 18.57 per cent of those interviewed said they didn't see any problems. Seven and a half per cent (7.40%) said that the lack of a sewage collection system is the main difficulty faced by the population. The other 7.40 per cent said that the way the sewage system is, it gets in the way of pedestrians and cars.

PHOTO 16 - Sewerage exposed to the open air
Source: JOECIO SANTOS SOUSA, 2013

PHOTO 16 shows a cave found in the São Pedro neighbourhood, where rainwater flows from the micro-drainage systems and sewage produced in the area. In addition, it was also possible to identify the accumulation of solid waste inside, further aggravating the situation regarding the healthiness of the environment in the region, which has several homes in its surroundings and on its banks.

TABLE 21 below deals with the consequences observed by residents in relation to the environmental aspect discussed above.

TABLE 21 - Observed consequences of the environmental aspect - sewage system - according to residents of the São Pedro neighbourhood

ENVIRONMENTAL IMPACT	OBSERVED CONSEQUENCE	PERCENTAGE THE ANSWERS
	Changes the domestic routine	11,11%
	Attracts insects and rodents	22,22%

Sewerage system	Exudes a bad odour	33,33%
	Diseases	29,62%
	Accumulation of rubbish	3,70%

As for the consequences observed in relation to the environmental aspect discussed in the table above, 33.33% of those interviewed said that the stench is the main nuisance, due to the exposure of open sewage. This was followed by 29.62% who said that illness was the main consequence, 22.22% who said that the sewage system attracts animals and insects, 11.11% who said that it changes their domestic routine and 3.70% who said that it causes rubbish to accumulate.

With regard to the environmental aspect, solid waste, the question was asked as to what kind of problems the population could observe in relation to waste management. TABLE 22 shows the information relevant to this issue.

TABLE 22 - Problems related to the environmental aspect - Solid Waste - according to residents of the São Pedro neighbourhood

ENVIRONMENTAL ASPECT	RELATED PROBLEMS	PERCENTAGE OF RESPONSES
Solid Waste	Insufficient collection	13,04%
	Proliferation of disease vectors	8,69%
	Improperly stored on the pavement	21,73%
	No problem at all	56,52%

According to the data shown in TABLE 22, 56.52% of those interviewed said they didn't see any problems related to solid waste in the neighbourhood, which is worrying given that, as identified in PHOTO 12, the neighbourhood has several problems that have taken on a negative weight, since 21.73% said that the main problem is improper storage on the pavement, 13.04% said that solid waste collection is still insufficient to meet the population's needs and 8.69% said that the proliferation of disease vectors was the most worrying problem. It's important to note that at no point did the interviewees mention the burning of solid waste as one of the problems, even though this situation occurs at various points in the neighbourhood (PHOTO 17).

PHOTO 17 - Burning solid waste in the São Pedro neighbourhood

Source: JOÉCIO SANTOS SOUSA, 2013

TABLE 23 below shows the consequences of the problems related to solid waste pointed out in the previous table, according to the opinion of the residents of the São Pedro neighbourhood.

TABLE 23 - Consequences observed on the environmental aspect - Solid Waste - according to residents of the São Pedro neighbourhood

ENVIRONMENTAL ASPECT	OBSERVED CONSEQUENCES	PERCENTAGE OF RESPONSES
Solid Waste	Diseases	17,64%
	Attracts animals	23,52%
	Bad odour	29,41%
	Impedes circulation	17,64%
	Visual pollution	11,76%

Among the consequences observed, 29.41 per cent said that bad odour was the main nuisance, 23.52 per cent said it was the ability to attract animals, 17.64 per cent said that solid waste hinders circulation and causes disease and 11.76 per cent said that the way it is distributed, it ends up causing visual discomfort (visual pollution).

At the time, residents of the São Pedro neighbourhood were also asked what kind of problems they could observe in relation to local afforestation and what the consequences of this were for their quality of life. Thus, TABLES 24 and 25 address the proposed problems.

TABLE 24 - Problems related to the environmental aspect - afforestation - according to residents of the São Pedro neighbourhood

ENVIRONMENTAL ASPECT	RELATED PROBLEMS	PERCENTAGE OF RESPONSES
Afforestation	Insufficient	55,55%
	Poorly cared for	7,40%
	Planted in inappropriate places	7,40%
	No problem at all	29,62%

TABLE 25 - Observed consequences of the environmental aspect - afforestation - according to residents of the São Pedro neighbourhood

ENVIRONMENTAL ASPECT	OBSERVED CONSEQUENCE	PERCENTAGE OF RESPONSES
Afforestation	Little shade	81,25%
	Makes it difficult for people to move around	6,25%
	Conflicts with infrastructure	12,5%

According to TABLE 24, 55.55% of residents said that the trees in the neighbourhood were insufficient, which explains why 81.25% of the answers converged on the option of little shade (TABLE 25), which in turn accentuates the population's feeling of heat. As for the other problems related to the neighbourhood's tree planting, 7.40% said that the trees were poorly looked after or planted in inappropriate places, while 29.62% said they didn't see any problems in this respect. As for the consequences observed, 6.25% said that the way they are arranged makes it difficult for people to move around and 12.5% said that the trees end up conflicting with the local infrastructure.

Thus, after checking all the information mentioned above, TABLE 26 assesses the environmental perception indicator, based on the percentage of responses obtained for each environmental aspect proposed in the tables above.

TABLE 26 - Assessment of the indicator - Environmental Perception

ITEM	INDICATOR	FEATURES	WEIGHTS
05	Environmental Perception	Perception of the environmental aspect - sewage system	+
		Perception of the environmental aspect - solid waste	-
		Perception of the environmental aspect - afforestation	+
TOTAL			+2 points - 1 point

Therefore, the environmental perception indicator, according to its assessment based on the environmental aspects addressed, obtained + 2 points from the sum of the two positive characteristics, as well as - 1 point due to the residents' perception of the environmental aspect of solid waste.

3.3 Comparison of the urban environmental quality of the neighbourhoods

In relation to all the indicators listed and assessed in the topics above, we can see situations in which the neighbourhoods scored both positively and negatively due to the characteristics presented in the field. TABLE 27 below compares the Vermelha and São Pedro neighbourhoods in terms of their scores and negative characteristics.

TABLE 27 - Comparison of negative scores for UQ indicators

NEIGHBOURHOODS	SCORES			
	QAU INDICATORS			
	Solid Waste	Cover Vegetable	Infrastructure and well-being	Perception Environmental
Red	-4	-1	- 1	-2
St Peter's	- 4	-1	-1	- 1

With regard to the information in TABLE 27, the sum of the negative characteristics of Vermelha resulted in - 8 points. In contrast, the São Pedro neighbourhood scored -7 points, one less than the former. Due to the similarities in characteristics, both neighbourhoods do not differ significantly in terms of negative points, since at various times they have had problems in common, such as inadequate solid waste disposal, a significant reduction in vegetation cover, deteriorated street structures and the way in which the population perceives the dynamics of the neighbourhoods.An important point to note in relation to the vegetation cover index of the neighbourhoods is that even though there has been a drop in green areas in the period between 2006 and 2010 in both cases, as seen in MAPS 04 and 05, and they are below the UN standards, there is nevertheless a difference in the percentage between the Vermelha and São Pedro neighbourhoods. TABLE 28 therefore shows a comparison of the neighbourhoods in terms of LCI.

TABLE 28 - Comparison of the ICV of the neighbourhoods

	RED	ST PETER'S	% IDEAL
ICV	17,50%	27,50%	30%

In terms of the ideal situation for a neighbourhood's vegetation cover, São Pedro is the closest to the 30% parameter. Therefore, specifically in this case, it has a favourable characteristic for its urban environmental quality.

Now, in terms of the positive characteristics assumed by the QAU indicators, the neighbourhoods had the following scores, which can be seen in TABLE 29.

TABLE 29 Comparison of the positive scores for the QAU indicators

NEIGHBOUR HOODS	QAU INDICATORS	SCORES		
	Solid Waste	**Vegetation cover**	**Infrastructure and well-being**	**Environmental Perception**
Red	+1	-	+9	+1
St Peter's	+3	-	+8	+2

Thus, TABLE 29 shows that the Vermelha neighbourhood scored +12 points and São Pedro scored +13 points. This also shows that the

scores due to the similar characteristics incorporated by the indicators in both neighbourhoods during the course of the research. Therefore, according to TABLE 30, the São Pedro and Vermelha neighbourhoods had the following Urban Environmental Quality Index.

TABLE 30 - IQAU presented by the Vermelha and São Pedro neighbourhoods

QUALI INDEX]	URBAN ENVIRONMENTAL DIMENSION
RED	**ST PETER'S**
Good	Good

In this way, both neighbourhoods had a good IQAU, according to the methodology applied, as the sum of the positive scores for the indicators were in the range of 10 - 15 points, which classifies them as neighbourhoods with good urban environmental quality.

CHAPTER 4

FINAL CONSIDERATIONS

The challenge of measuring urban environmental quality in a way that shows the real situation of a given location comes up against the very aspects that make up that reality, because the urban environment, for example, is made up of various factors resulting from human actions on its spaces, and the product of these actions directly interferes with the quality of life of the population and the physical environment.

In addition, each methodology has its own vision of how to study urban space, based on different indicators or parameters that are somehow related to the formation of a place's identity and especially its quality. In this way, this research used a simplified and experimental methodology to measure the urban environmental quality of the São Pedro and Vermelha neighbourhoods, using indicators that are part of the routine reality of both places.

After evaluating each indicator individually, it was decided that the neighbourhoods had both favourable and unfavourable situations, taking them into account when considering urban environmental quality.

Therefore, the Vermelha and São Pedro neighbourhoods are shown to have a good urban environmental quality index according to the methodology applied, due to the sum of all the positive scores presented by the indicators in the field. However, this result alone is not enough to claim that, in relation to all the aspects that make up the urban environment, it makes the neighbourhoods places with good urban environmental quality, or the opposite, as measurement approaches can vary the results.

In fact, this research also sought to correlate the population's perception of their responsibility to contribute to shaping the environmental and quality of life in their neighbourhoods, since human beings are the main agents in this context. And from this, to sensitise the community to making decisions, especially individual ones, which are extremely important for the group's quality of life.

Based on the application of this methodology, this research has come to the conclusion that the urban environmental quality of a given location, especially the aforementioned neighbourhoods, is fragile, and for this very reason it alerts society in all its segments that the construction of a cultural, social, environmental, city and urban identity must be supported by spatial planning, taking into account all the factors that make up the environment, in order to avoid generating conflicts that put human and natural safety at risk.

It is therefore necessary for new research to implement methodologies for measuring urban environmental quality in order to reach an ideal point in terms of techniques for understanding the environmental reality of cities and then formulating public policies that ensure the conservation and protection of life.

CHAPTER 5

REFERENCES

BRAGA, T. M; FREITAS, A.P.G; DUARTE, G.S; CAREPA-SOUSA, J. **Indices of municipal sustainability: the challenge of measurement**. Belo Horizonte: UFMG/Cedeplar, 2003.

CASSILHA, G. A; CASSILHA, S. A. **Urban planning and the environment**. Curitiba: IESDE BRASIL S.A, 2009.

CORREA, R.L. **O Espaço Urbano**. Editora Ática, Série Princípios, 3rd edition, n.174, 1995. p.1-16.

CUNHA, P.B. **Realidade socioeconômica e ambiental de um agrupamento de bairros da zona norte de Teresina, Piauí**. Thesis (doctorate in geography) - Universidade Estadual Paulista - UNESP, Rio Claro, 2010.

DIAS, F. A; GOMES, L. A; ALKMIM, J. K. **Evaluation of the urban environmental quality of the Ribeirão do Lipa Basin through indicators, Cuiabá/MT**. Revista Sociedade & Natureza, Uberlândia, Apr. 2011.

FAÇANHA, A. C. The urban evolution of Teresina: agents, processes and spatial forms in the city. 1998. Dissertation (Master's in Geography) - Federal University of Pernambuco. Recife, 1998a.

JOSÉ, C. **Teresina - process of structuring and urban expansion and its environmental influences in the South Zone.** Available at: < http://meuartigo.brasilescola.com/geografia/teresinaprocesso-estruturacao-expansao-urbana- suas-.htm >. Accessed on: 03 May 2013, 15:30:32

LIMA, V. **Analysis of urban environmental quality in Osvaldo Cruz/SP**. Master's dissertation - São Paulo State University, Faculty of Science and Technology. Presidente Prudente : [s.n], 2007. Available at: < http://www.athena.biblioteca.unesp.br/exlibris/bd/bpp/33004129042P3/2007/lima v me pru d.pdf >. Accessed on: 23 September 2012, 20:25:12

LOPES, A. F; MENESES, F. S; BARROS, C. A. R. **Origin and formation of the urban space of Teresina**. IV Research and Innovation Congress of the North and Northeast Technological Education Network - CONNEPI - Belém: Pará, 2009.

MACEDO, T.J.R; ROCHA, Y.T. **Qualidade ambiental urbana do bairro Jaguaribe, Município de Osasco, Estado de São Paulo, Brasil.** VI Latin American Seminar on Physical Geography II Ibero-American Seminar on Physical Geography University of Coimbra, May 2010

MARKONI, M. A; LAKATOS, E. M. **Metodología Científica** - 4. ed. - São Paulo: Atlas, 2004.

MELO, C. C. C. J; BRUNA, G. C. **Urban and regional development of Teresina, Piauí, Brazil and its importance in the current framework of influence in the Regional Urban Network in Brazil.** 1st Cape Verde Regional Development Congress, 2009.

MOURA, A. R. **Urban environmental quality in the Santa Cecília neighbourhood (centre of São Paulo/SP): A comparative and monitoring study of the years 1992 and 2008**. Dissertation (Master's in Geography) - Federal University of Paraná. Curitiba, 2010.

MOTA, S. **Urbanisation and the Environment**. Rio de Janeiro: ABES, 1999.

NETTO, M. P; SILVA, R. S. **Urban ecosystems: the potential of urban ecology in the development of sustainable cities.** IX ECOECO National Meeting. Brasilia, 2011.

OLIVEIRA, R. G. U. L; SANTOS, L. M. M; MACHADO, C. B; SANTOS, V. A. **The centre of Teresina: evaluation of the requalification programmes in the Teresina agenda 2015 plan.** III Urbicentro International Seminar - Salvador: Bahia, 2012.

TERESINA PROFILE: Economic, Social, Physical and Demographic. Raimundo Leôncio Ferraz Fortes, coord. Teresina: Municipal Secretariat for Economic Development and Tourism - SEMDEC. 2010. Available at: < http://www.teresina.pi.gov.br/portalpmt/orgao/SEMDEC/doc/20100709-336-1461-D.pdf>. Accessed on: 23 September 2012, 21:20:00

TERESINA DATA BANK PROJECT: Characterisation of the Municipality. Teresina City Hall: Municipal Planning and Coordination Secretariat - SEMPLAN. Teresina, 2011. Available at: < http://www.teresina-pi-gov.com.br/media/uploads/documento/2011/11/CARACTERISTICASGERAIS.pdf>. Accessed on: 23 September 2012, 21:37:13

SANTOS, C. D. **A formação e produção do espaço urbano: discussões preliminares acerca da importância das cidades médias para o crescimento da rede urbana brasileira.** Revista Brasileira de Gestão e Desenvolvimento Regional - v.5, n. 1, p. 77 - 190 - Taubaté: São Paulo, 2009.

VIANA, B. A. S. **The meaning of the city: between urban evolution and the process of verticalisation.** Carta Cepro, v. 23, n. 1, 2005.

APPENDIX I - FORM

1 - SOCIO-ECONOMIC PROFILE

A) WHAT IS THE SEX OF THE HEAD OF HOUSEHOLD?

1) male 2) female

B) WHAT IS THE EDUCATION LEVEL OF THE HEAD OF THE HOUSEHOLD?

1) illiterate 2) complete primary education
3) incomplete primary education 4) complete secondary education
5) incomplete secondary education 6) complete higher education
7) incomplete higher education

2 ENVIRONMENTAL PERCEPTION

A)WHAT DO YOU MEAN BY THE ENVIRONMENT?

B) WHAT DO YOU UNDERSTAND BY AN ENVIRONMENTAL PROBLEM?

C) DO YOU NOTICE ANY ENVIRONMENTAL PROBLEMS IN YOUR NEIGHBOURHOOD?

1) Yes 2) No
If so, which one(s)? ______________________________

D)IF IT EXISTS, HOW DOES IT AFFECT YOU?

E)IF NO, JUSTIFY IT!

2 - ENVIRONMENTAL ASPECTS OF THE NEIGHBOURHOOD AND HOME

A) WHAT KIND OF INFRASTRUCTURE DO YOU HAVE ACCESS TO?

1) sewerage 2) water supply 5) paving

3) rubbish collection 4) street lighting

3 - ENVIRONMENTAL PROBLEMS IN THE NEIGHBOURHOOD AND AT HOME

3.1 - SEWAGE SYSTEM

A) DO YOU NOTICE ANY PROBLEMS WITH THE SEWAGE SYSTEM IN YOUR NEIGHBOURHOOD?

1) exposed to open air 2) obstructs passage 3) accumulation of dirt
4) no problem at all 5)others ___________________________________

B) IF YES, WHAT CONSEQUENCE DO YOU SEE AS A RESULT?

1) changes the domestic routine 2) attracts insects and rodents
3) gives off a bad odour 4) diseases

3.2 - SOLID WASTE

A) DO YOU NOTICE ANY PROBLEMS WITH RUBBISH IN YOUR NEIGHBOURHOOD?

1) lack of collection 2) insufficient collection 3) proliferation of disease vectors
4) stored unduly on the pavement 5) there is no problem at all
6) other ____________________________

B) IF YES, WHAT CONSEQUENCE DO YOU SEE AS A RESULT?

1) diseases 2) attracts animals 3) bad odour

4) impedes traffic 5) visual pollution

3.3 - AFFORESTATION

A) ARE THERE ANY PROBLEMS WITH THE TREES IN YOUR NEIGHBOURHOOD?

1) insufficient 2) poorly cared for 3) planted in inappropriate places
4) no problem at all 5) others ____________________________

B) IF SO, WHAT CONSEQUENCES DO YOU SEE?

1) little shade 2) impedes the movement of people

3) conflicts with infrastructure 4) other ______________________

3.4 - NOISE POLLUTION

A) IN YOUR OPINION, IS THERE NOISE POLLUTION IN YOUR NEIGHBOURHOOD?

1) YES 2) NO

B) IF YES, WHAT TYPE OF NOISE POLLUTION?

1) traffic 2) neighbours 3) bar 4) car stereos 5) others __________________

C) WHAT CONSEQUENCE DO YOU SEE IN AS A RESULT?

1) prevents concentration 2) headache 3) disturbs sleep

4) disrupts studies 5) other ____________________

3.5 - AIR POLLUTION

A) IN YOUR OPINION, IS THERE AIR POLLUTION IN THE NEIGHBOURHOOD?

1) yes 2) no

B) IF YES, WHAT TYPE OF AIR POLLUTION?

1) emission of pollutants from vehicles 2) dust
3)soot 4) bad odour 5) no pollutants

C) WHAT DO YOU SEE AS THE MAIN CONSEQUENCE OF THIS?

1) nasal irritation 2) eye irritation 3) respiratory diseases
4) dirty the house 5) other ____________________

3.6 - URBAN DRAINAGE

A) DO YOU NOTICE WATER ACCUMULATING IN YOUR NEIGHBOURHOOD DURING THE RAINY SEASON?

1) yes 2) no

B) WHAT PROBLEM DO YOU SEE AS A RESULT?

1) accumulation of mud 2) accumulation of rubbish 3) risk of accidents
4) prevents passage 5) attracts insects

5) USES AND ORIGIN WITH QUALITY
THE URBAN AND DOMESTIC ENVIRONMENT

5.1 WATER

A) WHAT IS YOUR HOUSEHOLD'S WATER SUPPLY?

1) public network2) private source3) collective source

5.2 SOLID WASTE

A) HOW DO YOU STORE RUBBISH AT HOME?

1) closed container 2) open container 3) does not store at home 4) other

B) HOW DO YOU STORE RUBBISH OUTSIDE YOUR HOME?

1) closed container 2) open container
3) does not store outside 4) other

C) IS THERE REGULAR RUBBISH COLLECTION?

1) yes2) no

D) WHAT TIME OF DAY DO YOU PUT THE RUBBISH OUT?

1) morning2) afternoon 3) evening

4.3 AFFORESTATION

A) HOW YOU CONTRIBUTE TO CONSERVATION OF TREES IN YOUR NEIGHBOURHOOD?

1) pruning 2) watering 3) not damaging
4) planting5) has no interest in the subject

4.4 PROLIFERATION OF VECTORS

A) DO YOU CONTRIBUTE TO THE FIGHT AGAINST DENGUE?

1) yes 2) no
In what way? ____

MIX
Papier aus verantwortungsvollen Quellen
Paper from responsible sources
FSC® C105338

Printed by Books on Demand GmbH, Norderstedt / Germany